FIRE SATETY LOG BOOK

DATE (FROM)

CONTENTS OF THE LOG BOOK

1

GENERAL INFORMATION

AN INTRODUCTION TO YOUR LOG BOOK

The Fire Safety Order of 2005 says that the person in charge of a building must keep all fire safety tools and equipment working well and in good shape. If there are workers, they must be given proper training on how to stay safe in case of a fire.

This law also says that checks, upkeep, and training should be recorded so they can be reviewed to make sure they're happening.

This fire safety log book helps the person in charge keep track of these safety activities.

It helps show that you're following the fire safety laws by keeping records of fire safety measures. It's suggested to use this book in a way that you can add new pages when needed. Keep this book up-to-date and ready for the Fire Service to check if they need to.

It's illegal to write down something false on purpose.

Building Maintenance and Security	Fire fighting equipment maintenance and repairs
Electrical equipment test engineers	Emergency lighting maintenance and repairs
Environmental health departmen	Fire alarm maintenance and repairs
Health and safety executive	Fire safety advice and equipment supplies

LIST OF COMPETENT PERSONS AND FIRE WARDENS

Name	Department	Telephone	Ext.
Deputy			
Name	Department	Telephone	Ext.
Deputy			
Name	Department	Telephone	Ext.
Deputy			
Name	Department	Telephone	Ext.
Deputy			
Name	Department	Telephone	Ext.
Deputy			
Name	Department	Telephone	Ext.
Deputy			

FIRE SAFETY ADVICE

The tips below are to help you and your team stop fires from starting. If a fire does happen, they'll also help you avoid injuries or major damage to your building.

Escape Routes

- Make sure fire doors are always closed unless held open by a special device. Don't block them or mess with their self-closing features.
- Keep paths and stairs free of stuff that could get in the way. Make sure you can open the main exit door from the inside without a key, and keep the area outside this door clear.
- Mark exits and paths to them clearly, especially if they're not used often. Make sure signs pointing to the exits can be seen from far away in a room.

Fire Alarm

- Check the fire alarm system regularly to make sure it works. Teach your staff how to use it and what to do if it goes off.

Fire Extinguishers and Hose Reels

- These are for small, starting fires. Make sure everyone knows where they are and how to use them properly.
- Check and maintain them regularly. If you can check some yourself, follow the instructions and write down when you do it.

Lights
- Keep emergency and regular lights in working order. Fix or replace broken parts right away.

Staff and Guest Instructions

- Teach your staff what to do in an emergency, including:
    - How to sound the alarm.
    - How to call the fire department.
    - When to leave a fire alone.
    - How to use a fire extinguisher safely.
    - How to get out of the building correctly.
    - What's in the Fire Risk Assessment.

- Visit www.fireservices4u.co.uk for staff training courses.
- Make sure guests know what to do in an emergency too. If you have guests from other countries, provide fire instructions in their languages.

Electrical Safety

- Fires can start because of electrical problems. Check old wiring and update it if needed. If you're using more electrical devices, you might need more circuits. Always use the right fuse. Unplug appliances when you're not using them, especially overnight or when you leave.

Heating

- Keep the area around your boiler clear. Don't store things there. Keep things that can catch fire away from portable heaters.

Open Fires

- Never use flammable liquids to start a fire. Always protect fires with a guard.
- Clean your chimneys twice a year, or even more if you're burning wood.

Smoking

- Keep a close eye on areas where smoking is allowed and make sure there are enough ashtrays.
- Before leaving a room that won't be used for a while, or where people might sleep, double-check for any lit cigarette butts that could be hidden in furniture or bedding.
- Throw away ashtray contents into a bin that can't catch fire, and make sure everything is completely put out.

Be careful of common fire risks:

Electricity

Electricity can cause fires. If there's an electrical problem, get a skilled electrician to fix it right away. Remember to turn off gadgets when you're not using them.

Trash

Fires can easily start in trash. Throw away your garbage into metal bins with lids as soon as you can, and do it often. Keep bins away from your building to prevent fires from spreading.

Smoking

Cigarettes are a leading cause of fires. Be extra cautious.

Heaters

Be careful with portable heaters. They can cause fires if not used correctly or placed safely.

Dangerous Materials

Things like correction fluids, duplicator fluids, and aerosols can catch fire or explode. Keep them far from heat sources. Handling and storing any flammable liquids or gases safely is crucial.

Arson

To prevent intentional fires, lock up any flammable materials and secure your building, including doors and windows that might be overlooked.

If a fire happens

When the fire alarm goes off, leave the building using the planned escape route. Try to close doors and windows to slow down the fire. Call the Fire Brigade immediately, and have someone ready to guide them when they arrive.

DO NOT re-enter the building for any reason.

VISIT BY A FIRE BRIGADE OFFICER

Date	Inspecting Officer	Officers Signature	Comments

VISIT BY A FIRE BRIGADE OFFICER

Date	Inspecting Officer	Officers Signature	Comments

VISIT BY A FIRE BRIGADE OFFICER

Date	Inspecting Officer	Officers Signature	Comments

VISIT BY A FIRE BRIGADE OFFICER

Date	Inspecting Officer	Officers Signature	Comments

VISIT BY A FIRE BRIGADE OFFICER

Date	Inspecting Officer	Officers Signature	Comments

VISIT BY A FIRE BRIGADE OFFICER

Date	Inspecting Officer	Officers Signature	Comments

16

VISIT BY A FIRE BRIGADE OFFICER

Date	Inspecting Officer	Officers Signature	Comments

VISIT BY A FIRE BRIGADE OFFICER

Date	Inspecting Officer	Officers Signature	Comments

2

FIRE ALARM SYSTEM

FIRE ALARM SYSTEM

Fire alarm checks should be done the right way, following the maker's guide and the latest UK standards. It's key to test the fire alarm without causing a false fire alert.

Weekly Test by User:
- Every week, check the fire alarm works by testing a different manual call point (the spot where you can trigger the alarm manually). Do it at the same time each week. If your fire alarm is connected to a monitoring center, tell them before you start the test.

Quarterly Battery Inspection:
- Every three months, someone who knows about batteries should check vented batteries and their connections. Make sure the battery liquid (electrolyte) is at the right level and fill it up if needed.

Regular Checks by a Fire Alarm Engineer:
- A skilled fire alarm engineer should do more thorough checks and tests. How often these are needed depends on your alarm system's type and setup but usually happens every six months.

For Systems Without a Panel:
- If your fire alarm system doesn't have a control panel, press the test button on the alarm or manual call point to check it.

This routine ensures your fire alarm system is always ready to alert you in case of a fire.

FIRE DETECTORS

Regularly look over each smoke detector to make sure it's not damaged or covered in a lot of dirt, paint, or anything else that could stop it from working right. Each detector needs to be tested to see if it's working properly and is sensitive enough to detect smoke, just like the maker's guide and the latest UK rules say.

MEASURES TO REDUCE UNWANTED ALARMS

False alarms can interrupt work and be dangerous because they might send firefighters to the wrong place when they're needed elsewhere to save lives or property. To lower the chance of false alarms with automatic fire detectors, it's crucial to have a good system for testing and keeping them in shape. If a false alarm happens, it's important to figure out why and take steps to make sure it doesn't happen again.

AUTOMATIC DOOR RELEASE MECHANISM S ACTIVATED BY THE FIRE ALARM SYSTEM

Every week, when you test the fire alarm, also make sure that all the fire doors open and close properly into their frames. Remember to write down every check, test, and maintenance task you do, including any problems you find and fix. Also, note the date when you fix each issue.

FIRE ALARM TEST RECORDS

Date	Fire Alarm		Automatic Door Releases	Action Needed	Date Completed	Signature
	Location / Number	Satisfactory Yes / No	Satisfactory Yes / No			

FIRE ALARM TEST RECORDS

Date	Fire Alarm		Automatic Door Releases	Action Needed	Date Completed	Signature
	Location / Number	Satisfactory Yes / No	Satisfactory Yes / No			

FIRE ALARM TEST RECORDS

Date	Fire Alarm		Automatic Door Releases	Action Needed	Date Completed	Signature
	Location / Number	Satisfactory Yes / No	Satisfactory Yes / No			

FIRE ALARM TEST RECORDS

Date	Fire Alarm		Automatic Door Releases	Action Needed	Date Completed	Signature
	Location / Number	Satisfactory Yes / No	Satisfactory Yes / No			

FIRE ALARM TEST RECORDS

Date	Fire Alarm		Automatic Door Releases	Action Needed	Date Completed	Signature
	Location / Number	Satisfactory Yes / No	Satisfactory Yes / No			

FIRE ALARM TEST RECORDS

Date	Fire Alarm		Automatic Door Releases	Action Needed	Date Completed	Signature
	Location / Number	Satisfactory Yes / No	Satisfactory Yes / No			

FIRE ALARM TEST RECORDS

Date	Fire Alarm		Automatic Door Releases	Action Needed	Date Completed	Signature
	Location / Number	Satisfactory Yes / No	Satisfactory Yes / No			

FIRE ALARM TEST RECORDS

Date	Fire Alarm		Automatic Door Releases	Action Needed	Date Completed	Signature
	Location / Number	Satisfactory Yes / No	Satisfactory Yes / No			

FIRE ALARM TEST RECORDS

Date	Fire Alarm		Automatic Door Releases	Action Needed	Date Completed	Signature
	Location / Number	Satisfactory Yes / No	Satisfactory Yes / No			

FIRE ALARM TEST RECORDS

Date	Fire Alarm		Automatic Door Releases	Action Needed	Date Completed	Signature
	Location / Number	Satisfactory Yes / No	Satisfactory Yes / No			

FIRE ALARM TEST RECORDS

Date	Fire Alarm		Automatic Door Releases	Action Needed	Date Completed	Signature
	Location / Number	Satisfactory Yes / No	Satisfactory Yes / No			

FIRE ALARM TEST RECORDS

Date	Fire Alarm		Automatic Door Releases	Action Needed	Date Completed	Signature
	Location / Number	Satisfactory Yes / No	Satisfactory Yes / No			

FIRE ALARM TEST RECORDS

Date	Fire Alarm		Automatic Door Releases	Action Needed	Date Completed	Signature
	Location / Number	Satisfactory Yes / No	Satisfactory Yes / No			

FIRE ALARM TEST RECORDS

Date	Fire Alarm		Automatic Door Releases	Action Needed	Date Completed	Signature
	Location / Number	Satisfactory Yes / No	Satisfactory Yes / No			

3

FIRE
EXTINGUISHERS

FIRE EXTINGUISHER
INSPECTION AND MAINTENANCE

1. Regular Checks by You or Someone You Choose

You should regularly check all fire extinguishers, extra gas cartridges, and refill charges to make sure they are in the right spot, haven't been used, aren't losing pressure (if they have a pressure gauge), or haven't been damaged. These checks should happen at least every three months, but doing them every month is even better. If you find an extinguisher that can't be used, replace it right away.

2. Yearly Check-Up by a Professional

Make sure that a skilled professional inspects, services, and maintains your fire extinguishers, gas cartridges, and refill charges every year, following the latest British Standards. This person should know how to do these tasks well, using the right tools and following any specific instructions from the manufacturer. For extinguishers that you can check yourself, just look them over as the manufacturer suggests and write down what you find.

3. When to Empty and Refill

There are specific recommendations for how long you can go between emptying and refilling extinguishers, starting from either when they were made or the last time they were used or tested.

Type of extinguisher	Basic service	Extended service	Overhaul
Water-based	12-monthly	Every 5 years*	
Powder	12-monthly	Every 5 years*	
Powder-primary sealed	12-monthly	Every 5 years*	
Clean agent	12-monthly		Every 10 years
Halon	12-monthly		Every 10 years***
CO2	12-monthly		Every 10 years****

For water-based and powder fire extinguishers:
* You need to empty and refill them either 5 years after they start being used or 6 years after they were made, whichever date comes first. After that, they need to be serviced and refilled every 5 years from the date of the last service.

For powder fire extinguishers that are sealed and not meant to be refilled:
**They should be replaced 10 years after they start being used or 11 years after they were made, again, depending on which date comes first. After that, they need to be replaced every 10 years from the date of the last extended service.

*** For certain types of fire extinguishers, they can only be serviced for specific, important uses as outlined in Annex VII of EC Regulation 1005/2009.

For CO2 (Carbon Dioxide) extinguishers:

****The guidelines say you should base maintenance schedules on the date the extinguisher was made or last completely checked and serviced. This date is usually stamped on the extinguisher.

Remember, if you replace parts of the extinguisher, like the hose, it doesn't change the schedule for when the extinguisher needs to be emptied and checked. So, if you replace a hose on a CO2 extinguisher 6 years after it was new, you still need to test the extinguisher itself 4 years later.

For detailed guidelines on how and when to test fire extinguishers, look at the standards BS EN3 and BS 5306-3 Annex A & B.

FIRE EXTINGUISHERS - RECORD OF TESTS

Date	Result of Inspection	Remedial Action Taken	Fault Rectified (Date)	Signature

FIRE EXTINGUISHERS – RECORD OF TESTS

Date	Result of Inspection	Remedial Action Taken	Fault Rectified (Date)	Signature

FIRE EXTINGUISHERS – RECORD OF TESTS

Date	Result of Inspection	Remedial Action Taken	Fault Rectified (Date)	Signature

FIRE EXTINGUISHERS – RECORD OF TESTS

Date	Result of Inspection	Remedial Action Taken	Fault Rectified (Date)	Signature

FIRE EXTINGUISHERS – RECORD OF TESTS

Date	Result of Inspection	Remedial Action Taken	Fault Rectified (Date)	Signature

FIRE EXTINGUISHERS – RECORD OF TESTS

Date	Result of Inspection	Remedial Action Taken	Fault Rectified (Date)	Signature

FIRE EXTINGUISHERS – RECORD OF TESTS

Date	Result of Inspection	Remedial Action Taken	Fault Rectified (Date)	Signature

FIRE EXTINGUISHERS – RECORD OF TESTS

Date	Result of Inspection	Remedial Action Taken	Fault Rectified (Date)	Signature

FIRE EXTINGUISHERS – RECORD OF TESTS

Date	Result of Inspection	Remedial Action Taken	Fault Rectified (Date)	Signature

FIRE EXTINGUISHERS – RECORD OF TESTS

Date	Result of Inspection	Remedial Action Taken	Fault Rectified (Date)	Signature

FIRE EXTINGUISHERS – RECORD OF TESTS

Date	Result of Inspection	Remedial Action Taken	Fault Rectified (Date)	Signature

FIRE EXTINGUISHERS – RECORD OF TESTS

Date	Result of Inspection	Remedial Action Taken	Fault Rectified (Date)	Signature

4

FIRE HOSE
REELS

FIRE HOSE REEL
INSPECTION AND MAINTENANCE

The hose-reel test should be carried out in accordance with the manufacturer's instructions and the current British Standard

Annually - Each hose-reel should be completely run out and subjected to operational water pressure to ensure the hose is in good condition and that all couplings are water tight.

Note - All checks, tests and maintenance including faults and remedial action taken, should be recorded. The date each fault is rectified should also be recorded.

FIRE HOSE REELS – RECORD OF TESTS

Date	Hose Reel Location	Inspection / Test	Remedial Action Required	Date Completed	Name of Tester

FIRE HOSE REELS

Date	Hose Reel Location	Inspection / Test	Remedial Action Required	Date Completed	Name of Tester

FIRE HOSE REELS

Date	Hose Reel Location	Inspection / Test	Remedial Action Required	Date Completed	Name of Tester

FIRE HOSE REELS

Date	Hose Reel Location	Inspection / Test	Remedial Action Required	Date Completed	Name of Tester

FIRE HOSE REELS

Date	Hose Reel Location	Inspection / Test	Remedial Action Required	Date Completed	Name of Tester

FIRE HOSE REELS

Date	Hose Reel Location	Inspection / Test	Remedial Action Required	Date Completed	Name of Tester

FIRE HOSE REELS

Date	Hose Reel Location	Inspection / Test	Remedial Action Required	Date Completed	Name of Tester

FIRE HOSE REELS

Date	Hose Reel Location	Inspection / Test	Remedial Action Required	Date Completed	Name of Tester

FIRE HOSE REELS

Date	Hose Reel Location	Inspection / Test	Remedial Action Required	Date Completed	Name of Tester

FIRE HOSE REELS

Date	Hose Reel Location	Inspection / Test	Remedial Action Required	Date Completed	Name of Tester

FIRE HOSE REELS

Date	Hose Reel Location	Inspection / Test	Remedial Action Required	Date Completed	Name of Tester

FIRE HOSE REELS

Date	Hose Reel Location	Inspection / Test	Remedial Action Required	Date Completed	Name of Tester

5

EMERGENCY LIGHTENING

EMERGENCY LIGHTING

Emergency lighting tests need to follow the maker's guide and the latest UK standards.

Every Day: If your system uses a central power source, quickly check the indicator lights to make sure the system is ready to go.

Every Month: Pretend the main lights have gone out long enough to make sure all emergency lights turn on and work right. Look at each light to see if it's damaged or looks worn out, and check that the light covers and diffusers are clean.

Every Year: Fake a power cut to test if the emergency lights can last as long as they're supposed to on battery power. Also, check that the batteries are charging correctly.

Remember: Write down every check, test, and fix you do, including when you fix any problems.

EMERGENCY LIGHTING – RECORD OF TESTS

Date	Type of Test	Remedial Action Required	Date Completed	Name of Tester

EMERGENCY LIGHTING – RECORD OF TESTS

Date	Type of Test	Remedial Action Required	Date Completed	Name of Tester

EMERGENCY LIGHTING – RECORD OF TESTS

Date	Type of Test	Remedial Action Required	Date Completed	Name of Tester

EMERGENCY LIGHTING – RECORD OF TESTS

Date	Type of Test	Remedial Action Required	Date Completed	Name of Tester

EMERGENCY LIGHTING – RECORD OF TESTS

Date	Type of Test	Remedial Action Required	Date Completed	Name of Tester

EMERGENCY LIGHTING – RECORD OF TESTS

Date	Type of Test	Remedial Action Required	Date Completed	Name of Tester

EMERGENCY LIGHTING – RECORD OF TESTS

Date	Type of Test	Remedial Action Required	Date Completed	Name of Tester

EMERGENCY LIGHTING – RECORD OF TESTS

Date	Type of Test	Remedial Action Required	Date Completed	Name of Tester

EMERGENCY LIGHTING – RECORD OF TESTS

Date	Type of Test	Remedial Action Required	Date Completed	Name of Tester

EMERGENCY LIGHTING – RECORD OF TESTS

Date	Type of Test	Remedial Action Required	Date Completed	Name of Tester

MISCELLANEOUS TEST AND CHECKS

MISCELLANEOUS TEST AND CHECKS

As these systems are not found in the majority of premises this logbook only provides one page for recording the associated tests.
You should enter tests and results in this log book.

WEEKLY TESTS

For these systems, you usually need to keep a specific logbook. Ask your maintenance team or insurance provider for more details.

Sprinkler System Checks:
1. **Check Pressures and Water Levels:** Record the water and air pressure in the system, including the main pipes and pressure tanks. Also, check and note the water levels in any external water sources like reservoirs, rivers, lakes, or tanks.
2. **Test Water Motor Alarms:** Make sure each alarm sounds for at least 30 seconds.
3. **Check Engine Levels:** For diesel engines powering pumps, check the fuel and oil levels.
4. **Pump Tests:** Confirm automatic pumps start at the right water pressure. For diesel engines, check oil pressure, cooling water flow, or water levels in cooling systems, and try starting the engine with the manual button.
5. **Battery Maintenance:** Check the battery cells' electrolyte level and density. If the density is low, ensure the charging system works and replace any bad cells.
6. **Valve Monitoring:** Test the system that checks if stop valves are working in emergency systems.
7. **Alarm Connections:** Make sure the connection from the alarm switch to the control unit and from there to the Fire Service is working.
8. **Prevent Freezing:** Check that heating systems meant to stop the sprinkler system from freezing are working right.

Smoke Control Systems:
Simulate turning on the system to check if fans, exhaust ventilators, smoke dampers, and natural ventilators work as expected. Also, see if automatic smoke curtains move into place correctly.

MONTHLY TESTS

Smoke Control Systems for Helping Firefighters

Simulate turning on the system to check if the fans and exhaust systems work right. Make sure that smoke blockers (dampers) either shut or open properly, depending on what your system is supposed to do.

Monthly Checks and Tests

Every three months, have skilled people check and test the sprinkler system. Write down any problems they find and make sure they get fixed. Also, make sure you get a certificate that says the testing was done and everything's okay.

YEARLY TESTS

Yearly Checks and Fixes:

Make sure someone skilled does yearly checks on your sprinkler systems and smoke control systems. If they find any problems, write them down, fix them, and make sure to get a certificate saying everything was checked and is working fine.

ESCAPE ROUTE

Check escape routes and safety measures regularly to make sure nothing is blocking the way out. Doors should open easily, fire doors should close on their own, and fire-resistant walls and doors need to be in good shape. Also, all safety signs should be easy to read and in the right spots.

Write down every check, fix, and when each problem was solved.

GENERATORS

Always follow the maker's instructions. Sometimes engines don't start because of bad maintenance or problems with the battery or the start-up system. Dust and moisture can also cause issues, so keep everything clean and well-adjusted.

Make sure the generator's parts are secure, especially in places where there's a lot of shaking. Air going in and out of the generator shouldn't be blocked.

MISCELLANEOUS – RECORD OF TESTS

Date	Items Tested	Satisfactory Yes / No	Remedial Action	Signature

MISCELLANEOUS – RECORD OF TESTS

Date	Items Tested	Satisfactory Yes / No	Remedial Action	Signature

MISCELLANEOUS – RECORD OF TESTS

Date	Items Tested	Satisfactory Yes / No	Remedial Action	Signature

MISCELLANEOUS – RECORD OF TESTS

Date	Items Tested	Satisfactory Yes / No	Remedial Action	Signature

MISCELLANEOUS – RECORD OF TESTS

Date	Items Tested	Satisfactory Yes / No	Remedial Action	Signature

MISCELLANEOUS – RECORD OF TESTS

Date	Items Tested	Satisfactory Yes / No	Remedial Action	Signature

MISCELLANEOUS – RECORD OF TESTS

Date	Items Tested	Satisfactory Yes / No	Remedial Action	Signature

MISCELLANEOUS – RECORD OF TESTS

Date	Items Tested	Satisfactory Yes / No	Remedial Action	Signature

FIRE SAFETY
TRAINING & DRILLS

FIRE SAFETY TRAINING AND DRILLS FIRE SAFETY TRAINING

Fire Safety Training for Employees Should Cover:
- What to do if you find a fire.
- How to sound the alarm if there's a fire.
- What actions to take when the fire alarm goes off.
- Where to find fire extinguishers and how to use them, but only if it's safe.
- How to get out of the building safely.
- Where to go once you've evacuated (assembly points).
- How to alert the Fire and Rescue Service.
- How to help people who might need extra assistance during an evacuation.
- The risks of blocking fire exits or keeping fire doors open.

When to Train:
- When someone first starts working.
- If there are new or more dangers to know about.
- Regularly, at least once a year, depending on how risky the workplace is.

FIRE SAFETY DRILLS

Fire Drills:

- **Every 6 Months:** For places where people live, entertainment venues, big stores.

- **Every Year:** In industrial and commercial buildings.

Important Note:

These are the basic rules, but your workplace might need more frequent drills, like in care homes where it's important to practice at least twice a year to make sure everyone knows what to do.

RECORD OF FIRE SAFETY TRAINING & DRILLS

Date	Date of Appointment	Type of Training / Evacuation or Drill	Remedial Action	Signature

RECORD OF FIRE SAFETY TRAINING & DRILLS

Date	Date of Appointment	Type of Training / Evacuation or Drill	Remedial Action	Signature

RECORD OF FIRE SAFETY TRAINING & DRILLS CONT...

Date	Instruction	Person Receiving Instruction	Nature of Instruction	Signature

RECORD OF FIRE SAFETY TRAINING & DRILLS CONT...

Date	Instruction	Person Receiving Instruction	Nature of Instruction	Signature

Date	Instruction	Person Receiving Instruction	Nature of Instruction	Signature

RECORD OF FIRE SAFETY TRAINING & DRILLS CONT...

Date	Instruction	Person Receiving Instruction	Nature of Instruction	Signature

RECORD OF FIRE SAFETY TRAINING & DRILLS CONT...

Date	Instruction	Person Receiving Instruction	Nature of Instruction	Signature

RECORD OF FIRE SAFETY TRAINING & DRILLS CONT...

Date	Instruction	Person Receiving Instruction	Nature of Instruction	Signature

MISCELLANEOUS FIRE SAFETY CHECK LIST

MISCELLANEOUS FIRE SAFETY CHECK LIST

For Area _____ **Dated** _____

Completed by _____

Are the escape paths and hallways free from trash and things in the way?	☐ Yes	☐ No
Can all the emergency exit doors be opened easily without being locked?	☐ Yes	☐ No
Is all the equipment for fighting fires where it should be?	☐ Yes	☐ No
Are the fire safety signs easy to read and not damaged?	☐ Yes	☐ No
Can you see the evacuation instructions clearly in each room?	☐ Yes	☐ No
Do the fire doors shut all the way and tightly?	☐ Yes	☐ No
Are the fire doors kept closed and not blocked open?	☐ Yes	☐ No
Are all the computers turned off?	☐ Yes	☐ No
Does the electrical wiring look safe?	☐ Yes	☐ No
Has all the trash been taken out?	☐ Yes	☐ No

What action need to be taken

MISCELLANEOUS FIRE SAFETY CHECK LIST

For Area _____ **Dated** _____

Completed by _____

Are the escape paths and hallways free from trash and things in the way?	☐ Yes	☐ No
Can all the emergency exit doors be opened easily without being locked?	☐ Yes	☐ No
Is all the equipment for fighting fires where it should be?	☐ Yes	☐ No
Are the fire safety signs easy to read and not damaged?	☐ Yes	☐ No
Can you see the evacuation instructions clearly in each room?	☐ Yes	☐ No
Do the fire doors shut all the way and tightly?	☐ Yes	☐ No
Are the fire doors kept closed and not blocked open?	☐ Yes	☐ No
Are all the computers turned off?	☐ Yes	☐ No
Does the electrical wiring look safe?	☐ Yes	☐ No
Has all the trash been taken out?	☐ Yes	☐ No

What action need to be taken

MISCELLANEOUS FIRE SAFETY CHECK LIST

For Area _____ **Dated** _____

Completed by _____

Are the escape paths and hallways free from trash and things in the way?	☐ Yes	☐ No
Can all the emergency exit doors be opened easily without being locked?	☐ Yes	☐ No
Is all the equipment for fighting fires where it should be?	☐ Yes	☐ No
Are the fire safety signs easy to read and not damaged?	☐ Yes	☐ No
Can you see the evacuation instructions clearly in each room?	☐ Yes	☐ No
Do the fire doors shut all the way and tightly?	☐ Yes	☐ No
Are the fire doors kept closed and not blocked open?	☐ Yes	☐ No
Are all the computers turned off?	☐ Yes	☐ No
Does the electrical wiring look safe?	☐ Yes	☐ No
Has all the trash been taken out?	☐ Yes	☐ No

What action need to be taken

MISCELLANEOUS FIRE SAFETY CHECK LIST

For Area _____ **Dated** _____

Completed by _____

Are the escape paths and hallways free from trash and things in the way?	☐ Yes	☐ No
Can all the emergency exit doors be opened easily without being locked?	☐ Yes	☐ No
Is all the equipment for fighting fires where it should be?	☐ Yes	☐ No
Are the fire safety signs easy to read and not damaged?	☐ Yes	☐ No
Can you see the evacuation instructions clearly in each room?	☐ Yes	☐ No
Do the fire doors shut all the way and tightly?	☐ Yes	☐ No
Are the fire doors kept closed and not blocked open?	☐ Yes	☐ No
Are all the computers turned off?	☐ Yes	☐ No
Does the electrical wiring look safe?	☐ Yes	☐ No
Has all the trash been taken out?	☐ Yes	☐ No

What action need to be taken

MISCELLANEOUS FIRE SAFETY CHECK LIST

For Area _____ **Dated** _____

Completed by _____

Are the escape paths and hallways free from trash and things in the way?	☐ Yes	☐ No
Can all the emergency exit doors be opened easily without being locked?	☐ Yes	☐ No
Is all the equipment for fighting fires where it should be?	☐ Yes	☐ No
Are the fire safety signs easy to read and not damaged?	☐ Yes	☐ No
Can you see the evacuation instructions clearly in each room?	☐ Yes	☐ No
Do the fire doors shut all the way and tightly?	☐ Yes	☐ No
Are the fire doors kept closed and not blocked open?	☐ Yes	☐ No
Are all the computers turned off?	☐ Yes	☐ No
Does the electrical wiring look safe?	☐ Yes	☐ No
Has all the trash been taken out?	☐ Yes	☐ No

What action need to be taken

MISCELLANEOUS FIRE SAFETY CHECK LIST

For Area _____ **Dated** _____

Completed by _____

Are the escape paths and hallways free from trash and things in the way?	☐ Yes	☐ No
Can all the emergency exit doors be opened easily without being locked?	☐ Yes	☐ No
Is all the equipment for fighting fires where it should be?	☐ Yes	☐ No
Are the fire safety signs easy to read and not damaged?	☐ Yes	☐ No
Can you see the evacuation instructions clearly in each room?	☐ Yes	☐ No
Do the fire doors shut all the way and tightly?	☐ Yes	☐ No
Are the fire doors kept closed and not blocked open?	☐ Yes	☐ No
Are all the computers turned off?	☐ Yes	☐ No
Does the electrical wiring look safe?	☐ Yes	☐ No
Has all the trash been taken out?	☐ Yes	☐ No

What action need to be taken

MISCELLANEOUS FIRE SAFETY CHECK LIST

For Area _____ **Dated** _____

Completed by _____

Are the escape paths and hallways free from trash and things in the way?	☐ Yes	☐ No
Can all the emergency exit doors be opened easily without being locked?	☐ Yes	☐ No
Is all the equipment for fighting fires where it should be?	☐ Yes	☐ No
Are the fire safety signs easy to read and not damaged?	☐ Yes	☐ No
Can you see the evacuation instructions clearly in each room?	☐ Yes	☐ No
Do the fire doors shut all the way and tightly?	☐ Yes	☐ No
Are the fire doors kept closed and not blocked open?	☐ Yes	☐ No
Are all the computers turned off?	☐ Yes	☐ No
Does the electrical wiring look safe?	☐ Yes	☐ No
Has all the trash been taken out?	☐ Yes	☐ No

What action need to be taken

MISCELLANEOUS FIRE SAFETY CHECK LIST

For Area _____ **Dated** _____

Completed by _____

Are the escape paths and hallways free from trash and things in the way?	☐ Yes	☐ No
Can all the emergency exit doors be opened easily without being locked?	☐ Yes	☐ No
Is all the equipment for fighting fires where it should be?	☐ Yes	☐ No
Are the fire safety signs easy to read and not damaged?	☐ Yes	☐ No
Can you see the evacuation instructions clearly in each room?	☐ Yes	☐ No
Do the fire doors shut all the way and tightly?	☐ Yes	☐ No
Are the fire doors kept closed and not blocked open?	☐ Yes	☐ No
Are all the computers turned off?	☐ Yes	☐ No
Does the electrical wiring look safe?	☐ Yes	☐ No
Has all the trash been taken out?	☐ Yes	☐ No

What action need to be taken

MISCELLANEOUS FIRE SAFETY CHECK LIST

For Area _____ **Dated** _____

Completed by _____

Are the escape paths and hallways free from trash and things in the way?	☐ Yes	☐ No
Can all the emergency exit doors be opened easily without being locked?	☐ Yes	☐ No
Is all the equipment for fighting fires where it should be?	☐ Yes	☐ No
Are the fire safety signs easy to read and not damaged?	☐ Yes	☐ No
Can you see the evacuation instructions clearly in each room?	☐ Yes	☐ No
Do the fire doors shut all the way and tightly?	☐ Yes	☐ No
Are the fire doors kept closed and not blocked open?	☐ Yes	☐ No
Are all the computers turned off?	☐ Yes	☐ No
Does the electrical wiring look safe?	☐ Yes	☐ No
Has all the trash been taken out?	☐ Yes	☐ No

What action need to be taken